by Chenille Davis
illustrated by John Hovell

HMH

Copyright © by Houghton Mifflin Harcourt Publishing Company

All rights reserved. No part of this work may be reproduced or transmitted in any form or by any means, electronic or mechanical, including photocopying or recording, or by any information storage and retrieval system, without the prior written permission of the copyright owner unless such copying is expressly permitted by federal copyright law. Requests for permission to make copies of any part of the work should be submitted through our Permissions website at https://customercare.hmhco.com/contactus/Permissions.html or mailed to Houghton Mifflin Harcourt Publishing Company, Attn: Intellectual Property Licensing, 9400 Southpark Center Loop, Orlando, Florida 32819-8647.

Printed in Mexico

ISBN 978-1-328-77222-0

2 3 4 5 6 7 8 9 10 0908 25 24 23 22 21 20 19 18 17

4500675335   A B C D E F G

If you have received these materials as examination copies free of charge, Houghton Mifflin Harcourt Publishing Company retains title to the materials and they may not be resold. Resale of examination copies is strictly prohibited.

Possession of this publication in print format does not entitle users to convert this publication, or any portion of it, into electronic format.

Skip wants flowers.
He wants them for Flossie.

How many flowers does Skip have now?

Skip wants 9 flowers.
This one smells sweet.

How many flowers does Skip have now?

Skip looks for more.
He sees 3 flowers.

4   How many flowers does Skip have all together?

Skip likes red.
He finds 2 red flowers.

How many flowers will Skip have now?

5

Hooray! Here are 3 more.
Skip picks 3 white flowers.

6   How many flowers does Skip have in all?

Skip has 9 flowers.
Flowers for Flossie!

Responding

Math Concepts

Make a Bouquet

Draw Visualize

Draw a picture of a flower Skip picked.

Tell About
1. Look at page 5.
2. Tell someone a math story about how many flowers are in Skip's hand and how many more he found.

Write

Write how many flowers Skip picked for Flossie.